Searchlight
BOOKS™

What
Are Earth's
Cycles?

Investigating the

Carbon

Cycle

Mary Lindeen

Lerner Publications ◆ Minneapolis

Content Consultant: Katsumi Matsumoto, Professor of Earth Sciences, University of Minnesota

Lerner Publications Company
A division of Lerner Publishing Group, Inc.
241 First Avenue North
Minneapolis, MN 55401 USA

For reading levels and more information, look up this title at
www.lernerbooks.com.

Library of Congress Cataloging-in-Publication Data

Lindeen, Mary, author.
    Investigating the carbon cycle / by Mary Lindeen.
        pages cm. — (Searchlight books. What are earth's cycles?)
    Summary: "Eye-catching photos, informative captions, and succinct yet engaging text introduce young readers to the carbon cycle."— Provided by publisher.
    Audience: Ages 8–11
    Audience: Grades 4 to 6
    Includes index.
    ISBN 978-1-4677-8060-5 (lb : alk. paper) — ISBN 978-1-4677-8333-0 (pb : alk. paper) — ISBN 978-1-4677-8334-7 (eb pdf)
    1. Carbon cycle (Biogeochemistry)—Juvenile literature. I. Title. II. Series: Searchlight books. What are earth's cycles?
    QH344.L56 2016
    577.144—dc23   2015001947

Manufactured in the United States of America
1 – VP – 7/15/15

Contents

Chapter 1

CARBON IS EVERYWHERE

What do you have in common with a leaf, a cow, and a rock? You are all made from carbon. None of you would exist without the carbon cycle.

This bird and tree branch both contain carbon. What else contains carbon?

The Carbon Cycle

Carbon is everywhere. It is in the air. It is in the ocean. It is in plants and animals. It is in soil and rocks. Carbon moves between air, water, plants, animals, the ground, and machines. This movement is called the carbon cycle.

Most of Earth's carbon is in rocks and oceans. But carbon is also in plants, soil, and the atmosphere.

Carbon in Living Things

Plants that grow on land take in carbon from the air. Underwater plants take in carbon from the water. Plants use sunlight to turn carbon into their own food. A plant puts some of this carbon back into the air or water. Most of the carbon stays in the plant.

Plants put some of their carbon back into the atmosphere through a process called respiration.

LIONS, LIKE ALL ANIMALS, ARE PART OF THE CARBON CYCLE BECAUSE THEY EAT FOOD THAT CONTAINS CARBON.

All animals, including humans, need plants to live. Some animals eat plants. Some animals eat animals that eat plants. Other animals eat both plants and animals. When an animal eats, carbon that was in plants goes into that animal.

These tropical fish put carbon back into the ocean when water leaves their gills.

Cycling Carbon

Animals return carbon to the air or the ocean. Carbon goes into the air when land animals exhale. Ocean animals get rid of carbon in the water. Then plants can take in the carbon again.

All plants and animals are organic. That means they are living things, and all living things contain carbon. When organic things die, their remains start to decay. These remains can break down into tiny bits of matter. Carbon from plants and animals then goes into the soil and into water in the ocean.

THIS DEAD TREE HAS BEGUN TO DECAY. IN TIME, THE TREE'S CARBON WILL MAKE ITS WAY INTO THE SOIL.

This rock formation has many layers of sediment.

Carbon Underground

Organic remains do not always break down in soil. Sometimes layers of sediment cover the remains. The organic matter gets buried deeper and deeper. Pressure and heat inside Earth change the remains. This takes millions of years.

Buried organic matter still has carbon in it. After being buried for a long time, some of the organic matter turns into oil. Some of it becomes coal. And some of it turns into natural gas. Oil, coal, and natural gas have carbon in them.

A large truck carries coal away from a mine.

Carbon from Machines

People use oil, coal, and natural gas as fuel. Fuels from oil, coal, or natural gas are called fossil fuels. They come from animals and plants that lived long ago. Cars, furnaces, and other machines burn fossil fuels. When the fuel is burned, carbon from the fuel goes into the air. Then plants use the carbon again to make food. The carbon cycle continues.

Most cars burn fossil fuels, which puts more carbon into the air.

See the Cycle

People use fossil fuels every day in many ways.
Natural gas is often used for cooking and heating.
Many power plants burn coal to make electricity.
Gasoline, which comes from oil, powers cars. Try
to think of all the things you do in one day that
use fossil fuels. Make a list.

Chapter 2

WHAT IS CARBON?

We know carbon is all around us. But what exactly is carbon? Carbon is an element. Iron, gold, and helium are also elements. In fact, more than one hundred elements exist. Elements make up all the matter in the world.

Pennies contain copper, which is one of the many elements that exist in nature. What are some other elements?

An atom is the smallest possible unit of an element. Atoms are like building blocks. They join to form solids, liquids, or gases. When two or more atoms bond together, they form a molecule.

GOLD IS A SOLID AT ROOM TEMPERATURE, BUT IT BECOMES A LIQUID WHEN IT REACHES 1,948°F (1,064°C).

Bonding Time

Carbon atoms bond easily with other carbon atoms. Carbon bonds easily with other elements too. Oxygen is one of these elements. When two oxygen atoms bond with one carbon atom, a gas forms. This gas is carbon dioxide. It is the main form of carbon in the air.

Animals breathe out carbon dioxide. Plants take in carbon dioxide to make food. Carbon dioxide is also the gas made when we burn fossil fuels.

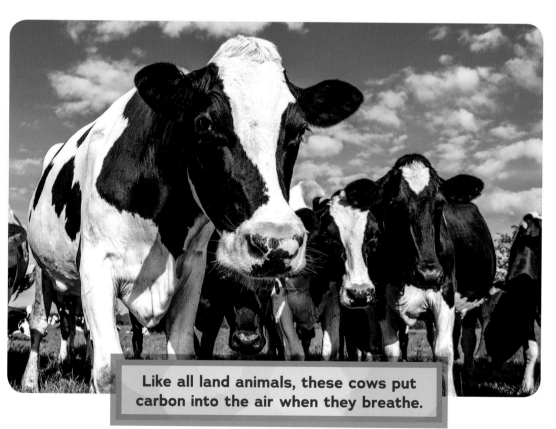

Like all land animals, these cows put carbon into the air when they breathe.

Diamonds form when carbon atoms bond in a certain pattern.

Carbon atoms can bond in different patterns. And different patterns make different materials. A pencil lead does not look like a diamond. But both are made of pure carbon. And both are made below Earth's surface. However, they are made at different pressures. Their carbon is bonded in different patterns too.

Under Pressure

It is very hot 100 miles (160 kilometers) underground. The carbon there is under great pressure. This heat and pressure causes the carbon atoms to form tight bonds. Each carbon atom bonds with four others. And each of those atoms bonds to four more. Billions of carbon atoms bonded in this pattern make a diamond. This is a very strong material.

DIAMOND MOLECULE

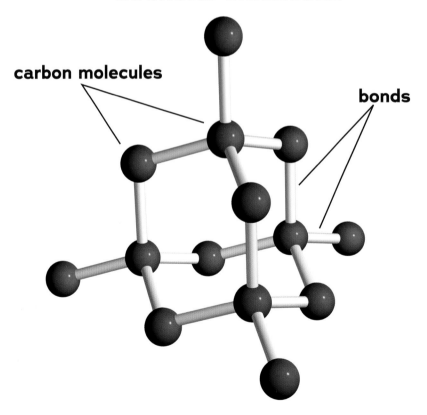

carbon molecules

bonds

See the Cycle

Put three toothpicks in the sides of a small marshmallow. Space the toothpicks evenly. Add one more toothpick on top of the marshmallow. Put a new marshmallow on the other end of each toothpick. Keep adding "atoms" in the same pattern. This is how carbon atoms bond in diamonds.

A pencil lead contains graphite. Graphite is solid carbon, like a diamond. But graphite forms in another way. It can come from spaces inside rocks. It can also come from coal. Rock and coal each contain carbon. Over time, heat and pressure inside Earth change the rock or the coal. When they break down, minerals separate. Carbon gets left behind in the form of graphite.

A miner holds a handful of coal.

GRAPHITE BREAKS APART AND LEAVES
A MARK WHEN YOU PRESS A PENCIL ON PAPER.

In graphite, the carbon atoms form rings. The bonds between atoms are weak. This makes graphite soft.

Chapter 3

WHY IS CARBON SO IMPORTANT?

You could live without a pencil or a diamond. But do you recall how you are like a leaf, a cow, and a rock? Without carbon, you would not exist. That is why carbon is so important to life on Earth.

These maple leaves contain carbon. Why is carbon so important to life on Earth?

The Backbone of Life

Your backbone supports your body. Carbon supports all life on Earth. In fact, carbon is sometimes called the backbone of life. Carbon bonds with many elements to make sugars and proteins. Plants and animals need these to live.

NONE OF THESE ANIMALS AND PLANTS WOULD EXIST WITHOUT CARBON.

Photosynthesis

Plants make food in their leaves. They do this through photosynthesis. Carbon dioxide and sunlight enter through a leaf. The plant takes in water through its roots. Tubes inside the stem bring the water to the leaves.

This tree is gathering the sun's energy so that it can make food.

Leaves then use the sun's energy to change the water and carbon dioxide. They are made into sugar and oxygen. The sugar gives the plant energy to grow. The plant saves some sugar for later. The plant lets go of the oxygen. Animals breathe in the oxygen that plants release.

PHOTOSYNTHESIS

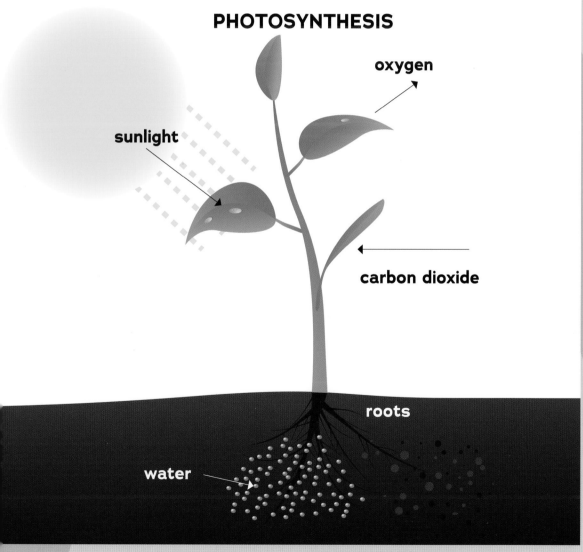

oxygen

sunlight

carbon dioxide

roots

water

A Warming Effect

Carbon dioxide in the air acts like a blanket around Earth. It traps heat and keeps us warm. Without carbon, the planet would freeze.

Nature keeps the carbon cycle in balance. But what if something changes that balance? When carbon dioxide builds up in our air, Earth's climate gets warmer than it used to be.

Power plants such as this one put extra carbon into the air.

See the Cycle

Crawl under a blanket. Tuck it all around you. Compare the temperature inside the blanket to the temperature outside the blanket. Add another blanket. The more blankets you add, the warmer it gets underneath. That is because the warm air you exhale and the heat from your body have a harder time escaping. Something similar happens when too much carbon dioxide builds up in the air that surrounds Earth. It traps more heat, and Earth gets warmer.

Chapter 4

CARBON AND CLIMATE CHANGE

Earth's climate has started to change. Temperatures around the world are getting warmer. Glaciers are melting. The water level in the oceans is rising. Trees are blossoming sooner in spring than they used to. These changes are affecting many plants and animals.

Average temperatures are increasing around the world. How are warmer temperatures affecting Earth?

Human activities are causing this change. People began using coal to run trains and ships about two hundred years ago. More machines were invented. They used more fuel. These days, electricity powers our homes, schools, and offices. Cars zip across town. Planes soar through the sky.

Machines make our lives easier. But many machines use fossil fuels. Burning more fossil fuels puts more carbon in the air. The amount of carbon dioxide we put in the air is called our carbon footprint. And the world's carbon footprint is getting bigger.

Planes help us travel faster, but they also put carbon into the air.

When people burn or cut down too many trees in an area, it is called deforestation. This is a major cause of climate change.

Too Much Carbon

In some places, extra carbon in the air helps trees grow. But in other places, people are burning and cutting down forests. When trees burn or decay, the carbon inside them is released. This adds more carbon dioxide to the air. And with fewer trees growing, there are fewer plants to use the carbon.

The ocean also absorbs carbon dioxide. But the extra carbon is changing the water. Many ocean animals struggle to survive. The warmer climate is also making glaciers melt. This adds water to the oceans. Changes are happening along coasts. Many animals no longer have the same habitats. Their food sources and homes are disappearing.

A LARGE PIECE OF ICE FALLS OFF A GLACIER.

Step by Step

Scientists have noticed these changes. People all around the world are taking action. Cities, states, and countries are making plans. For example, many cities are offering to pay part of the cost of installing solar panels on businesses and houses. Solar panels let us use power from the sun. These plans will help people use less fossil fuel. And this can reduce the amount of carbon going into the air.

People are also using new kinds of fuel. These fuels do not put as much carbon into the air. One new kind of fuel is made from vegetable oil or animal fat. Fuel can even be made with reused grease from cooking.

Scientists are worried about the amount of carbon in the air.

THE CARBON CYCLE

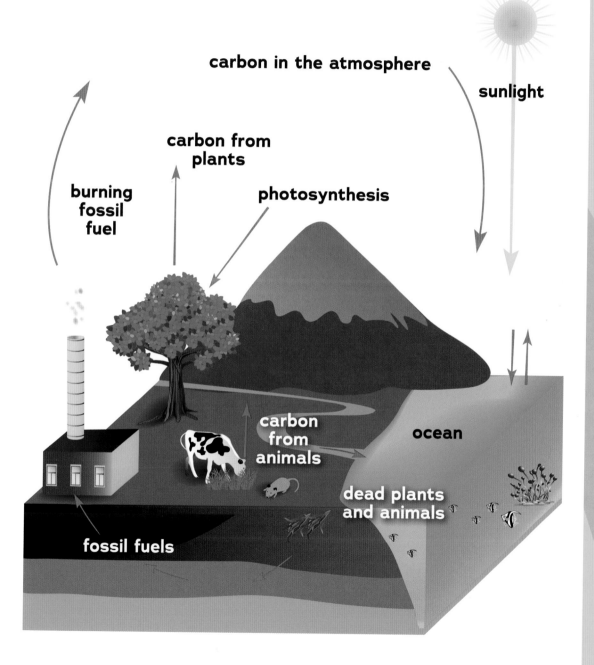

carbon in the atmosphere

sunlight

carbon from plants

photosynthesis

burning fossil fuel

carbon from animals

ocean

dead plants and animals

fossil fuels

These solar panels and wind turbines can create electricity without putting carbon into the air.

For many years, people have used fossil fuels to make electricity. But people are starting to use nature as energy. We can get electricity from wind and water. We can use the sun's energy too. Even the heat from below Earth's surface can be used to create energy.

See the Cycle

How can you use less fossil fuel? Walk or bike to school. Ask a parent if it's okay to turn the heat down when you sleep. Turn off lights when you leave a room. Unplug things that use electricity when you are not using them. Show your friends and family how to reduce their carbon footprints too!

It will take a worldwide effort to fight the effects of climate change. People will have to change their lifestyles and use fewer fossil fuels. Businesses will have to find ways to make money without harming the environment. Nations will have to pass laws to protect natural resources. Humans can reduce their carbon footprint. Understanding the carbon cycle is the first step to bringing it back into balance.

The US Congress has debated ideas for decreasing the amount of carbon in the air.

Science and the Carbon Cycle

Do this simple experiment to see how Earth's atmosphere traps heat. Put two thermometers next to each other on the ground. Face them both toward the sun. Record the temperature. Then put a clear glass bowl over one of the thermometers. Leave the other thermometer in the open air.

After fifteen minutes, record the temperature for each thermometer. Did the air under the bowl warm up faster than the air outside the bowl? The bowl is like the atmosphere surrounding Earth. It helps trap the warmth of the sun and makes the air surrounding Earth warmer.

Glossary

atom: the smallest possible piece of an element that still has all the properties of that element

bond: to join together

carbon: a chemical element found in many nonliving things, such as coal and diamonds, and in all plants and animals

carbon dioxide: a gas that is a mixture of carbon and oxygen

climate: the usual weather in a place

decay: to rot or break down

element: a substance that cannot be split into a simpler substance

fossil fuel: fuel that comes from oil, coal, or natural gas

mineral: a solid natural material that has a crystal structure

organic: of, relating to, or from living things

oxygen: a colorless gas found in the air

photosynthesis: a chemical process in which green plants make their food using water, carbon dioxide, and energy from the sun

sediment: solid pieces of material that are carried by water, wind, or ice from one place to another

Learn More about the Carbon Cycle

Books

Dakers, Diane. *The Carbon Cycle.* New York: Crabtree, 2015. Dakers explains how carbon moves through the planet's oceans, atmosphere, and living things.

Johnson, Rebecca L. *Investigating Climate Change: Scientists' Search for Answers in a Warming World.* Minneapolis: Twenty-First Century Books, 2009. In this title, readers discover how scientists investigate the causes and effects of climate change.

Mulder, Michelle. *Brilliant! Shining a Light on Sustainable Energy.* Victoria, BC: Orca, 2013. This informative book discusses unusual power sources that do not involve fossil fuels.

Websites

Changing the Balance: The Carbon Cycle
http://changingthebalance.thinkport.org/the_carbon_cycle.html
Interactive images guide readers through the various parts of the carbon cycle.

EPA: All about Carbon Dioxide
http://www.epa.gov/climatestudents/basics/today/carbon-dioxide.html
Visit this website to learn more about carbon and how it relates to climate change.

NASA: The Carbon Cycle
http://kids.earth.nasa.gov/seawifs/carbon.htm
This site provides many interesting facts to help readers understand the carbon cycle.

Index

Photo Acknowledgments

The images in this book are used with the permission of: © Erni/Shutterstock Images, p. 4; © Pavel Vakhrushev/Shutterstock Images, p. 5; © hraska/Shutterstock Images, p. 6; © Luisa Puccini/Shutterstock Images, p. 7; © Tischenko Irina/Shutterstock Images, p. 8; © suprunvitaly/iStockphoto, p. 9; © sigur/Shutterstock Images, p. 10; © abutyrin/Shutterstock Images, pp. 11, 20; © ssuaphotos/Shutterstock Images, p. 12; © Ulga/Shutterstock Images, p. 13; © Eldad Carin/Shutterstock Images, p. 14; © leomagdala/Shutterstock Images, p. 15; © stefbennett/Shutterstock Images, p. 16; © Jeffrey Hamilton/Digital Vision/Thinkstock, p. 17; © Vasilyev/Shutterstock Images, p. 18; © An Nguyen/Shutterstock Images, p. 19; © Purestock/Thinkstock, p. 21; © Shyshak Roman/Shutterstock Images, p. 22; © JREden/iStockphoto, p. 23; © vovan/Shutterstock Images, p. 24; © wawritto/Shutterstock Images, p. 25; © Gary Whitton/Shutterstock Images, p. 26; © wckiww/Shutterstock Images, p. 27; © Marccophoto/iStockphoto, p. 28; © MO_SES Premium/Shutterstock Images, p. 29; © Stockbyte/Thinkstock, p. 30; © Durk Talsma/Shutterstock Images, p. 31; © Monkey Business Images/Shutterstock Images, p. 32; © Designua/Shutterstock Images, p. 33; © Frank Fennema/Shutterstock Images, p. 34; © Thinkstock Images/Stockbyte/Thinkstock, p. 35; © Pablo Martinez Monsivais/AP Images, p. 36; © Evlakhov Valeriy/Shutterstock Images, p. 37.

Front cover: © Ken Gillespie Photography/Getty Images..

Main body text set in Adrianna Regular 14/20.
Typeface provided by Chank.